Tania Laura Barra Quispe
David Eleazar Barra Quispe
Paola Katherin Mantilla Cruz

NUTRITIONAL VALUE OF STREET FOODS

Tania Laura Barra Quispe
David Eleazar Barra Quispe
Paola Katherin Mantilla Cruz

NUTRITIONAL VALUE OF STREET FOODS

CALORIC INTAKE AND DOSAGE

ScienciaScripts

Imprint

Any brand names and product names mentioned in this book are subject to trademark, brand or patent protection and are trademarks or registered trademarks of their respective holders. The use of brand names, product names, common names, trade names, product descriptions etc. even without a particular marking in this work is in no way to be construed to mean that such names may be regarded as unrestricted in respect of trademark and brand protection legislation and could thus be used by anyone.

Cover image: www.ingimage.com

This book is a translation from the original published under ISBN 978-613-9-43341-4.

Publisher:
Sciencia Scripts
is a trademark of
Dodo Books Indian Ocean Ltd. and OmniScriptum S.R.L publishing group

120 High Road, East Finchley, London, N2 9ED, United Kingdom
Str. Armeneasca 28/1, office 1, Chisinau MD-2012, Republic of Moldova, Europe
Printed at: see last page
ISBN: 978-620-8-08902-3

INDEX

CALORIES OF RETAIL ARTISANAL FRUIT JUICES

Tania Laura Barra Quispe

National University of the Altiplano Puno

tanialbq@unap.edu.pe

Orcid: 0000-0003-1585-6314

950756197

Puno - Peru

David Eleazar Barra Quispe

National University of the Altiplano Puno

debarra@unap.edu.pe

Orcid: 0000-0003-0596-3829

Puno - Peru

Paola Katherin Mantilla Cruz

National University of the Altiplano

Orcid: 0000-0001-6996-5810

fabpaolamc7@gmail.com

Puno-Peru

SUMMARY

The nutrient composition of fruit juice is a key element to consider when analysing its influence on human health. This beverage, made by extracting fruit and consumed as a liquid, contains essential nutrients such as vitamins, minerals and antioxidants. The nutritional composition of fruit juice varies depending on the type of fruit. The aim of this study was to determine the caloric intake of special fruit juices sold to consumers in the city of Puno. The research was quantitative, observational, prospective, cross-sectional and descriptive in design. Sixty special juices from 60 juice shops located in four main markets in the city were analysed. The ingredients of each juice were weighed with a dietary scale, which made it possible to calculate the caloric and macronutrient intake. This procedure was carried out twice. It was found that the average macronutrient and kilocalorie content of a 280 ml serving of special fruit juice sold in the 60 juice shops in the four main markets of the city of Puno, Peru, was 143.6 g of carbohydrates, 14.0 g of fat, 20.6 g of protein, and a total of 783 kcal per serving. In conclusion, these juices are high in carbohydrates, mainly simple carbohydrates, compared to the other two macronutrients (fat and protein).

Keywords: Calorie intake, Artisanal fruit juice, Macronutrients

INTRODUCTION

It is vital to know how many calories are in the foods and drinks we consume in order to know how to work towards healthy eating habits.(Naomi et al. 2021). This acquired knowledge empowers us to be able to make informed choices about the foods we choose to eat, especially at a time when obesity, along with its related pathologies, is a worrying social issue (Basu and Penugonda 2009).(Basu and Penugonda 2009). In this context, it is crucial to investigate the calorie content of fruit juice, a very common beverage that is commonly and erroneously considered as healthy(Scheffers et al. 2022).. Beyond the pleasure that fruit juice can give us, it has some disadvantages that we should not turn our backs on. Important nutrients may be lost in the course of its production, which means that certain essential nutrients are no longer provided to the body. Furthermore, given the current state of sugar consumption, excessive consumption of fruit juice can lead to what is known as obesity, complications such as diabetes and heart disease (Choo et al. 2018).(Choo et al. 2018). Nowadays, therefore, it makes sense to be aware of the calories in our food and drinks(Rodriguez Delgado et al. 2017). A closer look at this aspect will allow us to be more attentive to our calorie intake and at the same time provide us with interesting information on the nutritional aspects of fruit juice (Ashraf et al. 2024).(Ashraf et al. 2024). From the detailed analysis of its nutrients, of the specific components from which it is made, we can make considered determinations as to whether or not we can incorporate the consumption of fruit juice into our diet(Naomi et al. 2021). The study of the calories in fruit juice, in the hands of the consumer, provides the consumer with the information necessary to understand more about the nutritional properties of the product (Lee et al. 2022). (Lee et al. 2022). This knowledge is not only useful for those who choose to replace their dietary habits with better ones, but can help to reinforce public policies to control the marketing of food products, especially those aimed at children. With measures to limit excessive energy consumption and the promotion of more nutritious alternatives, we predict and prevent diseases in future generations.(Basu and Penugonda 2009).

The importance of the nutritional and energy value of juices sold in "juice shops" is not only positive for making dietary choices, but also influences public health and the development of policy regulation (Melo and Peraçoli 2007).(Melo and Peraçoli 2007).. A thorough understanding of the true energy content of fruit juice can help everyone (all citizens) to make more informed choices and promote, more broadly and collectively, healthier dietary habits for the population (Ashraf et al. 2024). (Ashraf et al. 2024).

METHODOLOGY

The present study is a quantitative, observational, prospective, cross-sectional and descriptive design. Fifteen juice shops were selected from each of the main markets in the city of Puno (4 markets), for a total of 60 juice shops. In each establishment a ration of "special" juice was purchased, which is the most requested by consumers and whose price ranges between 7.00 s/. and 12.00s/. nuevos soles, depending on each establishment.

Before processing, each ingredient of the special juice was ordered, i.e. one by one, in special airtight jars. This made it possible to weigh them with the aid of a Soehnle dietary scale. In this way, the net weights used in the special fruit juices sold by the juice shops in the four markets were obtained. It is important to note that this procedure was carried out twice in order to be as close to reality as possible.

Using the net weights obtained, macronutrient and kilocalorie content was calculated using Nutricalcs software. These data were recorded and analysed in an Excel spreadsheet to obtain averages of kilocalories, carbohydrates, fats and proteins for each juice shop and market stall. These results were then presented in the form of graphs.

RESULTS

The macronutrient and kilocalorie content of a 280 ml serving of special fruit juice sold in the juice shops of the four main markets in the city of Puno, Peru, was identified. As shown in Figure 1, on average, the special juice from juice shops located in the main market 1 provides 174.9 g of carbohydrates and 699.7 kcal from this macronutrient, followed by the main market 2, whose content is 161.5 g of carbohydrates, equivalent to 646.1 kcal. Similarly, in terms of fat intake, as shown in figure 2, it is again the main markets 1 and 2 whose juice shops offer the special juice with the highest content of this macronutrient, being 18.7 g (167.9 kcal) and 15.1 g (135 kcal) of fat, respectively. In the same vein, in terms of protein intake (see figure 3), these two markets also offer the special fruit juice with the highest protein content, with the main market 1 providing 174.9 g (699 kcal) of protein and the main market 2, 161.5 g (646.1 kcal) of protein. Finally, on average (see figure 4), from the 60 juice shops located in the four main markets of this city, a serving of special fruit juice (280 ml), made from papaya, banana, pear, apple, pineapple, milk, carrot juice, egg, maca, carob, honey, pollen, nuts, cereals, beer and, in some cases, wine, provides an average of 143.6 g of carbohydrates, mainly simple carbohydrates, 14.0 g of fat, from both animal and vegetable sources, 20.6 g of protein, mainly of vegetable and to a lesser extent animal origin, and a total of 783 kcal from the whole preparation.

On the other hand, Figure 5 shows that the main market N°04 has the highest average cost of artisanal fruit juices (10.00 s/.), and at the same time, it has the highest caloric intake compared to the juice shops of market N°01, whose average price is 8.00 s/. and whose caloric intake is the lowest. However, market N°03 presents the lowest cost, being the second in the order of highest caloric contribution.

Table N°01 shows the variability in nutritional content and price of artisanal fruit juices in different markets, indicating both minimum and maximum values for each parameter. Main Market 01 has the highest variability in all aspects, while Main Market 02 has fixed values for price and lower variability in macronutrient and calorie content.

Figure 1. Carbohydrate content of special fruit juice sold to Puneño consumers.

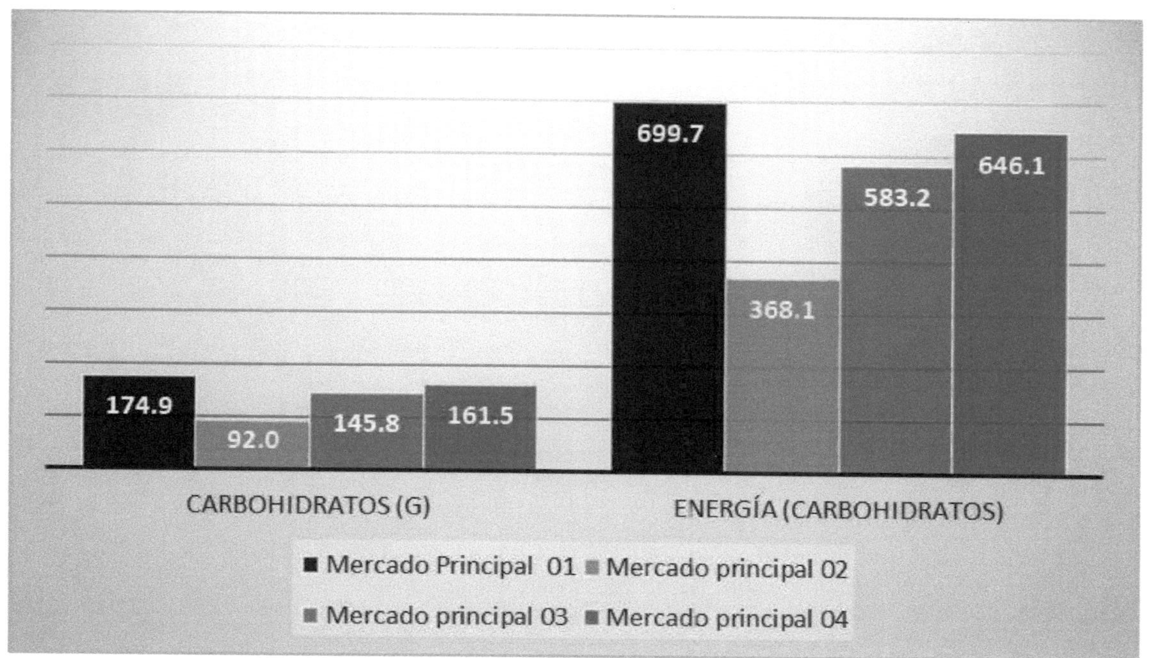

Figure 2. Fat content of special fruit juice sold to Puneño consumers.

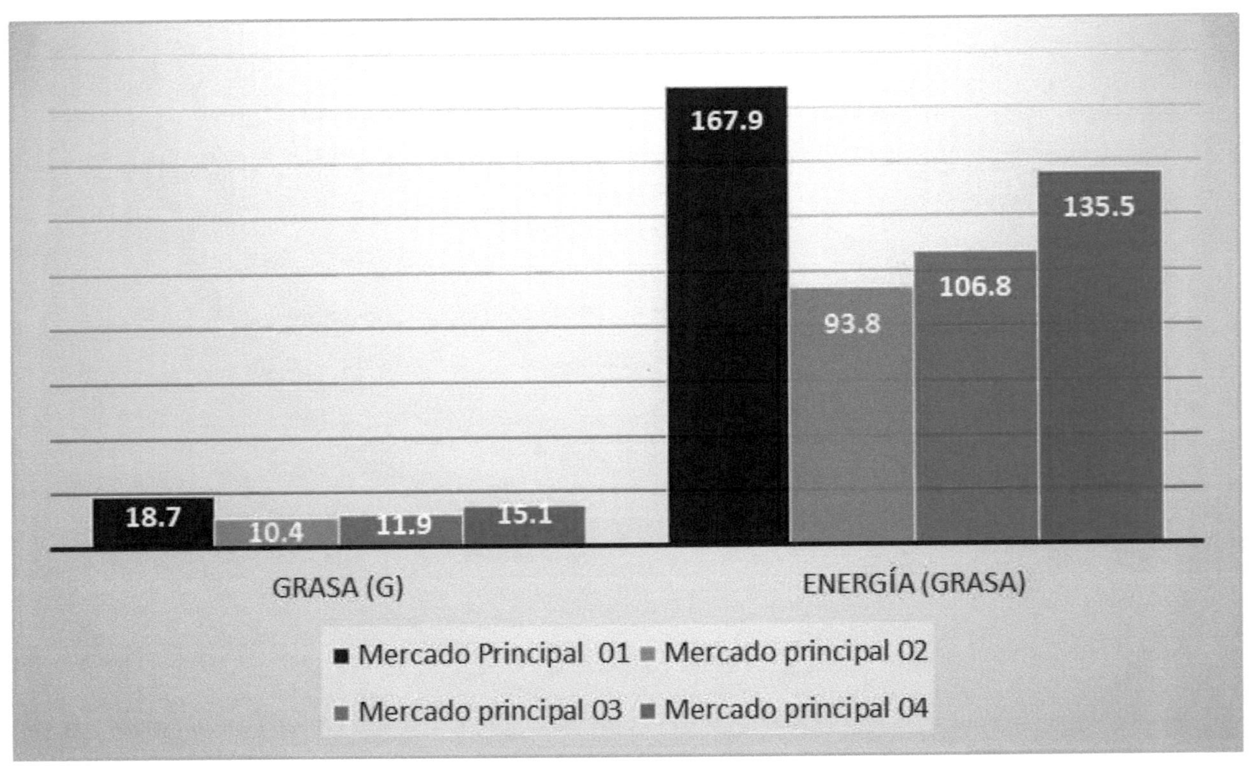

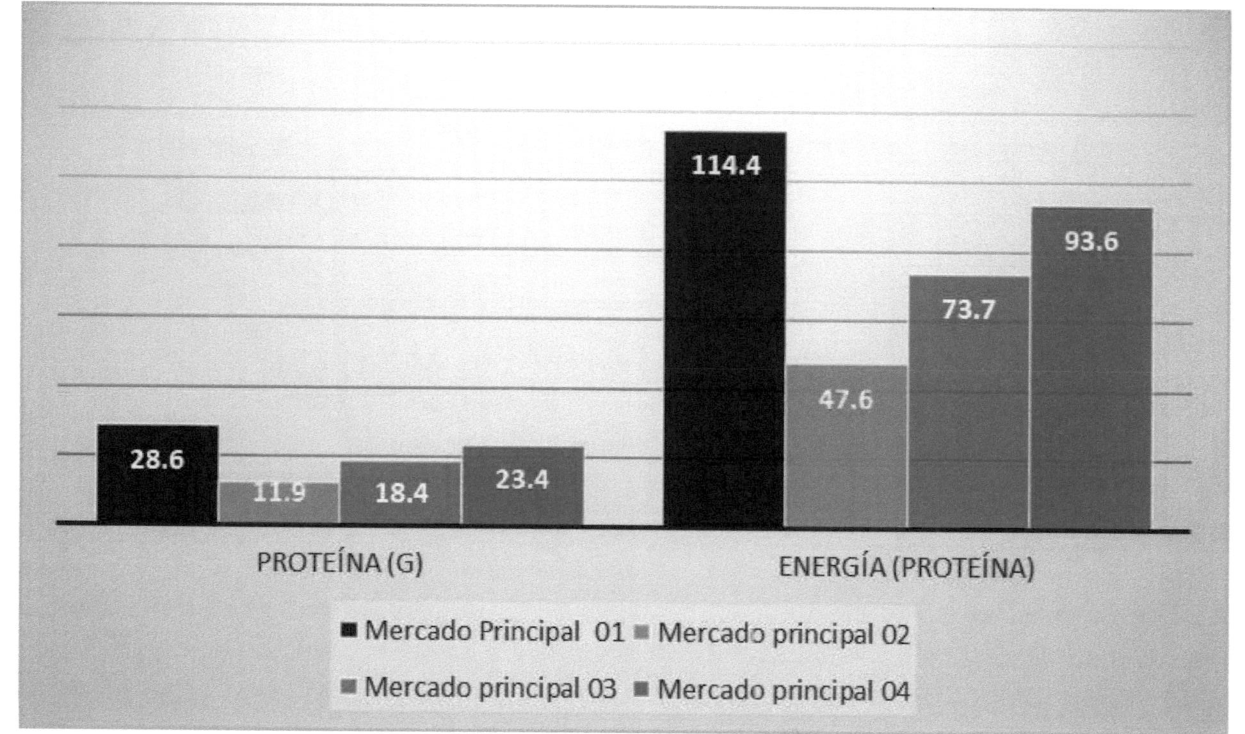

Figure 3. Protein content of special fruit juice sold to Puneño consumers.

Figure 4. Average macronutrient content of special fruit juice sold to Puneño consumers.

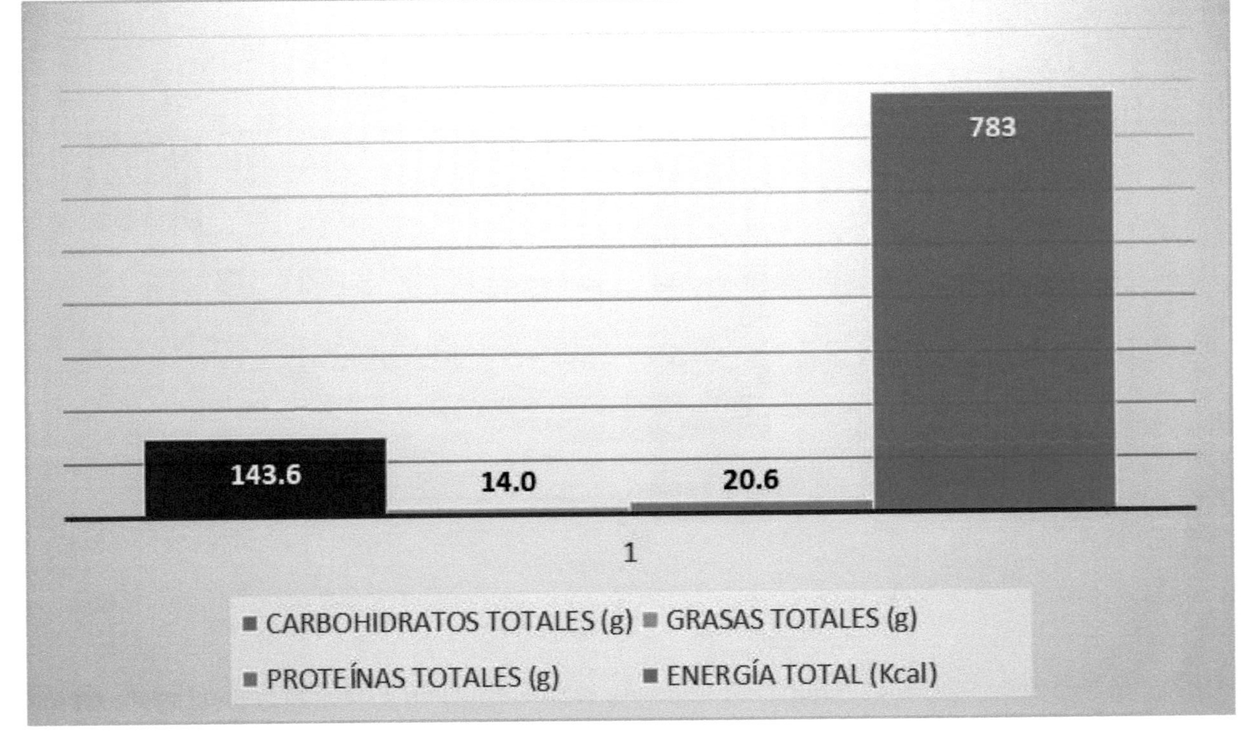

Figure 5. Average cost and content of macronutrients and total calories in Puno consumer special fruit juice.

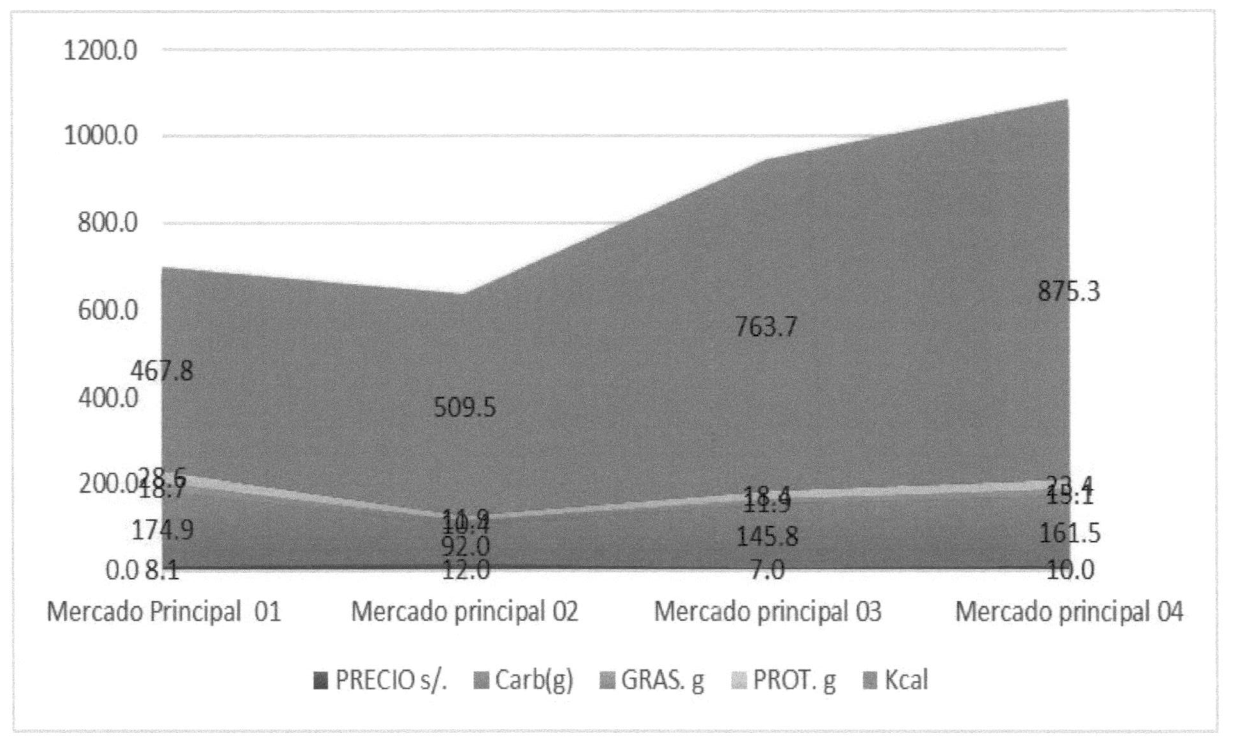

Table 1. Minimum and maximum macronutrient and calorie intake in Puno consumer speciality fruit juice

MARKET		PRICE s/.	CARB. g	GRAS. g	PROT. g	Kcal
Main Market 01	min	7	80.93	10.6	12.16	106.87
	max	11	492.18	28.48	51.78	557.31
Main Market 02	min	12	81.81	6.28	9.46	98.53
	max	12	132.98	17.53	14.81	150.61
Main Market 03	min	7	109.62	8.33	10.05	143.18
	max	7	228.26	15.38	27.25	263.84
Main Market 04	min	10	135.75	7.12	19.32	162.19
	max	10	207.81	24.133	30.193	262.136

DISCUSSION

In recent years, the consumption of fruit juice has experienced a considerable boom, as it could have an important impact on human health.(Tojo Sierra 2003). The speed with which we go through our daily routine and with which we usually eat breakfast, as we have to be on time for work, university or school, favours a breakfast that is sometimes quick and without appetite(Choo et al. 2018). The popular belief that skipping breakfast is detrimental to health, as it is the most important meal of the day, makes us choose to consume fruit juice as a quick option. However, this option is not always the healthiest one(J. Yu et al. 2023)Fruit juices can be very sugary and are often accompanied by products such as butter, margarine, jam, bologna, etc., which increase their calorie content and thus reduce their nutritional value. In fact, if it seems that fruit juice and bread with accompaniments are energy products for breakfast, what they actually have in common is that they are mainly made up of simple carbohydrates(Scheffers et al. 2022).. Public health recommendations indicate that free sugars, other than those in foods and beverages themselves, should not exceed 10% of daily calorie intake. In fact, for best health performance, it is generally advised not to exceed 5% of the daily intake of free sugars(Morales-Cahuancama et al. 2022)..

In the present study, it is possible to see how the homemade fruit juices sold in the juice shops of the city of Puno have a high contribution in kilocalories, with simple carbohydrates being the largest component (143.6 g), which exceeds the amount of carbohydrates needed for breakfast, considering a 25% of the total, with an equal distribution of 2000 kcal. If the habit of drinking these juices were to become established among the inhabitants of Puno, it could be detrimental to their health by increasing the risk of fat accumulation in the liver, a pathology known as non-alcoholic fatty liver disease (B. Yu et al. 20).(B. Yu et al. 2023)..

The increasing occurrence of fatty liver disease in the population has led to a growing interest in elucidating whether there is a link between fruit juice intake and the origin of this disease.(Naomi et al. 2022). Fatty liver disease has been gaining importance globally as a major public health problem, and research on the possible role of fruit juice intake in the development of this disease has intensified (J. Yu et al. 2022).(J. Yu et al. 2023)..

To understand the impact of fruit juice intake on health, it is necessary to look at the context of fruit juice intake, considering the general diet followed, with the influence of advertising and marketing.(Liu et al. 2023). The broad implications of fruit juice intake on the overall diet and the health outcomes it may have need to be considered in order to make an informed decision about the appropriateness of fruit juice intake (Jardí et al. 2019).(Jardí et al. 2019).

CONCLUSION

The results of this study conclude that a glass of approximately 280 ml of "special" artisanal fruit juice, made mainly with papaya, banana, pear, apple, pineapple, milk, carrot juice, egg, maca, carob, honey, pollen, nuts, 7 cereals, beer and wine, sold in the 60 juice shops in the 4 main markets of the city of Puno, provides an average of 783 kcal. Of these, 73% come from simple carbohydrates, 16% from fats and 10% from proteins.

REFERENCES

Ashraf, R., A. M. Duncan, G. Darlington, A. C. Buchholz, J. Haines, D. W. L. Ma, and Guelph Family Health Study the. 2024. "The degree of food processing contributes to sugar intakes in families with preschool-aged children". *Clinical Nutrition ESPEN* 59:37-47. doi: 10.1016/j.clnesp.2023.11.010.

Basu, Arpita, and Kavitha Penugonda. 2009. "Pomegranate Juice: A Heart-Healthy Fruit Juice." *Nutrition Reviews* 67(1):49-56. doi: 10.1111/j.1753-4887.2008.00133.x.

Choo, Vivian L., Effie Viguiliouk, Sonia Blanco Mejia, Adrian I. Cozma, Tauseef A. Khan, Vanessa Ha, Thomas M. S. Wolever, Lawrence A. Leiter, Vladimir Vuksan, Cyril W. C. Kendall, Russell J. de Souza, David J. A. Jenkins, and John L. Sievenpiper. 2018. "Food Sources of Fructose-Containing Sugars and Glycaemic Control: Systematic Review and Meta-Analysis of Controlled Intervention Studies." *BMJ (Clinical Research Ed.)* 363:k4644. doi: 10.1136/bmj.k4644.

Jardí, Cristina, Núria Aranda, Cristina Bedmar, Blanca Ribot, Irene Elias, Estefania Aparicio, and Victoria Arija. 2019. "Free sugars intake and overweight at early ages. Longitudinal study." *Anales de Pediatría* 90(3):165-72. doi: 10.1016/j.anpedi.2018.03.018.

Lee, Danielle, Laura Chiavaroli, Sabrina Ayoub-Charette, Tauseef A. Khan, Andreea Zurbau, Fei Au-Yeung, Annette Cheung, Qi Liu, Xinye Qi, Amna Ahmed, Vivian L. Choo, Sonia Blanco Mejia, Vasanti S. Malik, Ahmed El-Sohemy, Russell J. de Souza, Thomas M. S. Wolever, Lawrence A. Leiter, Cyril W. C. Kendall, David J. A. Jenkins, and John L. Sievenpiper. Leiter, Cyril W. C. Kendall, David J. A. Jenkins, and John L. Sievenpiper. 2022. "Important Food

Sources of Fructose-Containing Sugars and Non-Alcoholic Fatty Liver Disease: A Systematic Review and Meta-Analysis of Controlled Trials." *Nutrients* 14(14):2846. doi: 10.3390/nu14142846.

Liu, Q., L. Chiavaroli, S. Ayoub-Charette, A. Ahmed, T. A. Khan, F. Au-Yeung, D. Lee, A. Cheung, A. Zurbau, V. L. Choo, S. B. Mejia, R. J. de Souza, T. M. S. Wolever, L. A. Leiter, C. W. C. Kendall, D. J. A. Jenkins, and J. L. Sievenpiper. 2023. "Fructose-Containing Food Sources and Blood Pressure: A Systematic Review and Meta-Analysis of Controlled Feeding Trials." *PLoS ONE* 18(8 August). doi: 10.1371/journal.pone.0264802.

Mclo, Célia Regina Maganha e, and José Carlos Peraçoli. 2007. "Mensuração da energia despendida na fasting e ao inporte calorico (MIEL) em parturientas". *Revista Latino-Americana de Enfermagem* 15:612-17. doi: 10.1590/S0104-11692007000400014.

Morales-Cahuancama, Bladimir, Gandy Dolores-Maldonado, Paul Hinojosa-Mamani, William Bautista-Olortegui, Cinthia Quispe-Gala, Lucio Huamán-Espino, and Juan Pablo Aparco. 2022. "Analysis of the distribution of macronutrients in food baskets delivered by municipalities during the COVID-19 pandemic in Peru". *Revista Peruana de Medicina Experimental y Salud Pública* 39:6-14. doi: 10.17843/rpmesp.2022.391.9742.

Naomi, Novita, Elske Brouwer-Brolsma, Marion Buso, Sabita Soedamah-Muthu, Johanna Geleijnse, Anne Raben, Jo Harrold, Jason Halford, and Edith Feskens. 2021. "Sugar-Sweetened Beverages, Fruit Juice, and Low-Calorie Beverages, and All-Cause Mortality Risk Among Dutch Adults: The Lifelines Cohort Study Within the SWEET

Project." *Current Developments in Nutrition* 5:1066. doi: 10.1093/cdn/nzab053_059.

Naomi, Novita, Joy Ngo, Elske M. Brouwer-Brolsma, Marion E. C. Buso, Sabita S. Soedamah-Muthu, Carmen Pérez-Rodrigo, Anne Raben, Joanne A. Harrold, Jason C. G. Halford, Lluis Serra-Majem, Johanna M. Geleijnse, and Edith J. M. Feskens. 2022. "Association of Sugar-Sweetened Beverages, Low/No-Calorie Beverages and Fruit Juice Intakes with Non-alcoholic Fatty Liver Disease: The SWEET Project". *Current Developments in Nutrition* 6:934. doi: 10.1093/cdn/nzac067.054.

Rodríguez Delgado, J., M. S. Hoyos Vázquez, J. Rodríguez Delgado, and M. S. Hoyos Vázquez. 2017. "Fruit juices and their role in children's diets: should we consider them as just another sugary drink? Positioning of the Gastroenterology and Nutrition Group of the AEPap". *Pediatría Atención Primaria* 19(75):103-16.

Scheffers, Floor R., Jolanda M. A. Boer, Ulrike Gehring, Gerard H. Koppelman, Judith Vonk, Henriëtte A. Smit, W. M. Monique Verschuren, and Alet H. Wijga. 2022. "The association of pure fruit juice, sugar-sweetened beverages and fruit consumption with asthma prevalence in adolescents growing up from 11 to 20 years: The PIAMA birth cohort study". *Preventive Medicine Reports* 28:101877. doi: 10.1016/j.pmedr.2022.101877.

Tojo Sierra, R. 2003. "Consumption of fruit juices and beverages by Spanish children and teenagers: health implications of their poor use and abuse. *Anales de Pediatría* 58(6):584-93. doi: 10.1016/S1695-4033(03)78126-0.

Yu, Bowei, Ying Sun, Yuying Wang, Bin Wang, Xiao Tan, Yingli Lu, Kun Zhang, and Ningjian Wang. 2023. "Associations of artificially sweetened beverages, sugar-sweetened beverages, and pure fruit/vegetable juice with visceral adipose tissue mass". *Diabetes & Metabolic Syndrome: Clinical Research & Reviews* 17(10):102871. doi: 10.1016/j.dsx.2023.102871.

Yu, J., A. Mahajan, G. Darlington, A. C. Buchholz, A. M. Duncan, J. Haines, D. W. L. Ma, and Family Health Study Guelph. 2023. "Free sugar intake from snacks and beverages in Canadian preschool- and toddler-aged children: a cross-sectional study." *BMC Nutrition* 9(1). doi: 10.1186/s40795-023-00702-3.

QUANTITY OF INGREDIENTS DOSED IN THE MENUS SOLD IN RESTAURANTS IN THE CITY OF PUNO

Tania Laura Barra Quispe[1]

tanialbq@unap.edu.pe

https://orcid.org/0000-0003-1585-6314

National University of the Altiplano

Peru

David Eleazar Barra Quispe[2]

debarra@unap.edu.pe

https://orcid.org/0000-0003-0596-3829

National University of the Altiplano

Peru

Diamilet Rojas Mamani[3]

diamiletrojas@gmail.com

https://orcid.org/0009-0001-4474-9115

National University of the Altiplano

Peru

Juan Reynaldo Pardes Quispe [4]

jparedes@unap.edu.pe

https://orcid.org/0000-0002-3847-0554

National University of the Altiplano

Peru

SUMMARY

The dosage of ingredients in food preparation is a fundamental factor in obtaining satisfactory results. Ensuring the correct proportion of each ingredient contributes to the quality of the final dish. However, it is essential to understand the importance of this dosage to avoid problems such as inconsistent flavours and textures, risks of food poisoning among others. The objective of the study was to identify the dosage of food in food dishes sold to consumers in high traffic areas in the city of Puno, Peru. A descriptive, observational, prospective and cross-sectional study was carried out. The sample consisted of 44 commercial restaurants, from which 132 preparations were collected (44 breakfasts, 44 lunches and 44 dinners), which were taken to the laboratory of the School of Human Nutrition of the National University of the Alliplano for direct weighing and to obtain the net cooked weight, which was then converted to net raw weight using conversion factors, in order to contrast it with the recommended dosage in the table of food dosage for collective food services. As a result, it was found that the food dosages used in the 44 breakfasts, 44 lunches and 44 dinners did not conform to the recommendations set out in the food dosage table. It concluded that there were excesses in the cereal and tuber groups, while deficiencies were detected in the dosage of vegetables, meat and fish, and no fruit was included in the three meal times.

Keywords: Ingredient dosage, Menu, Puno dishes, Restaurant.

INTRODUCTION

Today, it is crucial to investigate the food served in restaurants, as the proliferation of restaurants worldwide is the result of changes in eating habits over time.(Villegas Villegas 2011). This topic has gained relevance due to the concern for people's health and well-being, the assessment of the nutritional quality of food, and the promotion of sustainable practices in the food industry. It also provides knowledge on the ingredients used, the identification of potential allergens and the prevention of food-related diseases. Food dosage is essential to ensure an adequate and balanced diet. Proper dosage ensures that people consume the right amount of nutrients, avoiding both under- and over-intake(Zheng et al. 2023). It contributes to maintaining health and preventing diet-related diseases, optimising the use of food resources and ensuring equitable access to food for the entire population. Proper dosage has a number of benefits for health and well-being. It allows nutritional needs to be met accurately, strengthening the immune system, preventing disease, maintaining a healthy weight, and controlling hunger and satiety.(Hernánzez Elizondo et al. 2019)..

Knowing the quantity of ingredients dosed in the preparations cooked by street restaurants is essential to ensure the quality of the food offered to consumers. This involves the proper use of each ingredient, prevents health problems, ensures the nutritional value of the food, and optimises production costs by avoiding waste and using ingredients efficiently.(Singh et al. 2023).

METHODOLOGY

The research is descriptive, observational, prospective and cross-sectional. A total of 132 meals, including 44 breakfasts, 44 lunches and 44 dinners, were collected from a sample of 44 food outlets. These venues were selected from a total of 50 using a mathematical equation for the calculation of sample size for a finite population. The selected establishments are commercial restaurants that offer a conventional menu with a price of no more than 10.00 nuevos soles, have infrastructure and are located in areas close to the university area, downtown area and market area of the city of Puno. Preparations were selected that allow the ingredients to be separated and that are commonly requested by consumers, excluding those places that offer preparations such as milk chocolate, api, quinoa juice, cañihua, salchipapas, grilled chicken, broaster chicken and similar.

The 44 restaurants were randomly selected by probability sampling using the online application "App Sorteos". Each preparation was taken to the laboratory of the School of Human Nutrition of the Universidad Nacional del Altiplano for direct weighing, a technique that consisted of separating the ingredients one by one using kitchen utensils such as plates, spoons, forks, strainers and measuring jugs. The net cooked weight was determined using a Soehnle digital dietary scale with a capacity of 5000g and an accuracy of 1g. This weight was then converted to raw net weight using the conversion factors from the corresponding table of cooked to raw food weight conversion factors. This allowed the over- or under-dosing to be checked against the food dosing table for mass catering.

The data obtained were processed in an Excel spreadsheet to calculate the average dosage of each ingredient in each preparation and meal time. The analysis included the determination of averages, and the results were presented in tables.

Deficiencies have been detected in the dosage of ingredients present in broths, soups and creams sold during breakfast in restaurants in the city of Puno, as detailed in table 1. Specifically, dosage deficiencies of more than 10 g were observed in ingredients such as pore (D at 11 g), cabbage (D at 21 g), onion (D at 23 g), beef (D at 65 g), chicken (D at 60 g) and chickpea (D at 24 g). On the other hand, an excess of more than 20 g was found in the dosage of rice (E at 124 g), rice (E at 67 g), moron (E at 60 g), cracked wheat (E at 148 g), potato (E at 20 g) and chuño (E at 22 g).

As for the stews and seconds sold during breakfast, as detailed in table 2, a deficient dosage of more than 10 g was observed in peas (D at 11 g), tomato (D at 34 g), vanita (D in 12 g), beef (D in 76 g), lamb (D in 95 g), beef part rib (D in 61 g), chicken (D in 58 g), potato (D in 29 g), omelette or fried egg (D in 25 g) and trout (D in 43 g). On the other hand, an excess of more than 20 g was found in the dosage of rice (E at 104 g), browned or fried potato (E at 31 g) and fried banana (E at 38 g).

In relation to broths, soups and creams served at lunchtime, as shown in table 3, deficiencies of more than 10 g were observed in the dosage of spinach (D at 18 g), paprika (D at 22 g), cabbage (D at 20 g), vanita (D at 18 g), beef (D at 65 g), chicken (D at 70 g), cassava (D at 24 g) and chickpea (D at 24 g), while rice (E at 128 g), noodles (E at 122 g), moron (E at 75 g) and quinoa (E at 162 g) were overdosed by more than 20 g.

Likewise, in relation to the stews and seconds served by the restaurants during lunch, as shown in Table 4, deficiencies of more than 10 g were observed in the dosage of broccoli (D at 16 g), paprika (D at 28 g), tomato (D at 35 g), beef (D at 75 g), pork (D at 80 g), chicken (D at 53 g), sausage (D at 33 g), liver (D at 23 g), horse mackerel (D at 75 g), omelette or fried egg (D at 49 g), maize (D at 33 g), cheese (D at 32 g), potato (D at 21 g), golden or fried potato (D at 22 g) and olive (D at 20 g). While those exceeding 20 g were rice (E at 84 g), fideo (E at 129 g), chuño (E at 11 g), sweet potato (E at 75 g) and ocopa (E at 32 g).

Regarding the broths, soups and creams served during dinner, as shown in table 5, there were deficiencies of 10 g in the dosage of spinach (D at 27 g), paprika (D at 23 g), cabbage (D at 15 g), vanita (D at 17 g), corn (D at 24 g), beef (D at 59 g), chicken (D at 41 g) and chuño (D at 33 g), maize (D at 24 g), beef (D at 59 g), chicken (D at 41 g) and chuño (D at 33 g), while rice (E at 117 g), noodles (E at 155 g), black pudding (E at 141 g) and semolina (E at 184 g) were exceeded by 20 g in the dosage.

Finally, in relation to the stews and seconds served by the restaurants during dinner, as shown in table 6, deficiencies of 10 g or more were observed for onion (D at 13 g), lettuce (D at 11 g), tomato (D at 33 g), beef (D at 71 g), chicken (D at 53 g), canned tuna steak (D at 110 g), potato (D at 31 g) and chicken (D at 53 g), tomato (D at 33 g), beef (D at 71 g), chicken (D at 53 g), canned tuna steak (D at 110 g), potato (D at 31 g) and lentil (D at 14 g), while those over 20 g were rice (E at 68 g) and noodle (E at 30 g).

Table 1

Dosage of ingredients in broths, soups and creams served in restaurants during breakfast

Food group	Ingredients used	PPNE	PPNR
	Celery	9g	16g
	Parsley	9g	15g
	Pore	4g	15g
Vegetables	Cabbage	3g	24g
	Onion	1g	24g
	Carrot	7g	15g
	Pumpkin	9g	21g
	Meat (beef)	5g	70g
	Meat (lamb)	84g	70g
Meat	Leg (lamb)	114g	120g
	Leg (beef)	138g	120g
	Chicken	10g	70g
Cereal grains	Rice	144g	20g

		PPNE	PPNR
	Rice	87g	20g
	Morón	90g	30g
	Broken wheat	168g	20g
Pulses and pulses and their derivatives	Chickpea	1g	25g
	Potato (cooked)	70g	50g
Root vegetables and preparations	Chuño	67g	45g

PPNE: Average net weight found, PPNR: Average recommended net weight

Table 2

Dosage of ingredients of casseroles and main courses served in restaurants at breakfast

Food group	Ingredients used	PPNE	PPNR
Vegetables	Peas	4g	15g
	Onion	41g	45g
	Lettuce	22g	30g
	Tomato	6g	40g
	Vanita	8g	20g
	Carrot	12g	20g
Meat	Meat (beef)	44g	120g
	Meat (lamb)	25g	120g
	Beef (rib)	59g	120g
	Chicken	62g	120
	Sausage	40g	40g
Cereal grains	Rice	204g	100g
	Noodle	91g	90g
Tubers and preparations	Pope	61g	90g

	Chuño	52g	50g
	Browned or fried potato	101g	70g
Milk and milk products	Cheese	42g	50g
Eggs	Eggs ((omelette, fried)	35g	60g
Fish	Trout	87g	130g
Fruits	Banana (fried)	68g	30g

PPNE: Average net weight found, PPNR: Average recommended net weight

Table 3

Dosage of ingredients in broths, soups and creams served in restaurants at lunchtime

Food group	Ingredients used	PPNE	PPNR
Vegetables	Celery	5g	16g
	Onion	1g	15g
	Spinach	2g	20g
	Haba	5g	15g
	Turnip	2g	15g
	Paprika	3g	25g
	Pore	1g	15g
	Cabbage	4g	24g
	Vanita	2g	20g
	Carrot	9g	15g
	Pumpkin	6g	21g
Cereal grains	Rice	148g	20g
	Noodle	142g	20g
	Morón	105g	30g

		PPNE	PPNR
	Quinoa	192g	30g
Meat	Meat (beef)	5g	70g
	Chicken	21g	70g
Root vegetables and preparations	Potato (cooked)	52g	50g
	Chuño	37g	45g
	Cassava	6g	30g
Legumes	Chickpea	1g	25g

PPNE: Average net weight found, PPNR: Average recommended net weight

Table 4

Dosage of ingredients of stews and main courses served in restaurants at lunchtime

Food group	Ingredients used	PPNE	PPNR
	Peas	3g	15g
	Betarraga	34g	30g
	Broccoli	14g	30g
	Onion	12g	25g
Vegetables	Lettuce	17g	30g
	Paprika	2g	30g
	Tomato	5g	40g
	Vanita	13g	25g
	Carrot	17g	20g
	Pumpkin	100g	95g
	Meat (beef)	45g	120g
Meat and meat preparations	Meat (pork)	40g	120g
	Chicken	67g	120g
	Sausage	7g	40g

Category	Food	PPNE	PPNR
	Liver	67g	90g
Fish	Horse mackerel	55g	130g
Eggs	Egg (omelette, fried)	11g	60g
	Rice	184g	100g
	Corn	36g	38g
Cereals, grains and derivatives	Noodle	219g	90g
	Maize	5g	38g
	Lentil	84g	60g
Pulses, legumes and derivatives	Pallar	29g	25g
Milk and dairy products	Cheese	18g	50g
	Potato (cooked)	69g	90g
	Potato (browned, fried)	48g	70g
Root vegetables and preparations	Chuño	61g	50g
	Sweet potato	145g	70g
Sauces and creams	Ocopa	92g	60g
Fruit and preparations	Olive	5g	25g

PPNE: Average net weight found, PPNR: Average recommended net weight

Table 5

Dosage of ingredients of broths, soups and creams served in restaurants at dinner time

Food group	Ingredients used	PPNE	PPNR
Vegetables	Celery	6g	16g
	Spinach	3g	30g
	Paprika	2g	25g
	Pore	2g	15g
	Cabbage	9g	24g
	Vanita	3g	20g
	Carrot	9g	15g
	Pumpkin	4g	21g
	Rice	137g	20g
	Noodle	175g	20g
Cereals, grains and derivatives	Maize	1g	25g
	Morón	171g	30g
	Semolina	214g	30g

		PPNE	PPNR
Meat	Meat (beef)	11g	70g
	Chicken	29g	70g
Tubers and tuber products	Potato (cooked)	40g	50g
	Chuño	12g	45g

PPNE: Average net weight found, PPNR: Average recommended net weight

Table 6

Dosage of ingredients of stews and main courses served in restaurants at dinner time

Food group	Ingredients used	PPNE	PPNR
	Peas	8g	15g
	Betarraga	18g	30g
	Onion	12g	25g
Vegetables	Lettuce	19g	30g
	Tomato	7g	40g
	Vanita	10g	15g
	Carrot	12g	20g
Cereals, grains and derivatives	Rice	168g	100g
	Noodle	130g	100g
Meat	Meat (beef)	49g	120g
	Chicken	67g	120g
Fish and fish by-products	Tuna fillet (canned)	14g	124g
Tuber roots and derivatives	Potato (cooked)	59g	90g
	Potato (browned, fried)	73g	70g

		52g	50g
	Chuño	52g	50g
	Olluco	104g	100g
Leguminous plants and their derivatives	Lentil	46g	60g

PPNE: Average net weight found, PPNR: Average recommended net weight

DISCUSSION

The need to eat away from home has become a daily activity due to various factors such as access to food, academic commitments, lack of time and current work demands.(Omoniyi and Cosmas 2024).. This triggers a number of consequences on a person's energy balance. Therefore, the health professional will always recommend eating the preparations at home, as this ensures that the ingredients used and the approximate quantities are correct (Redondo Del Río et al.(Redondo Del Río et al. 2016) which makes it possible to choose healthier options and avoid those that may be harmful to health. This is not the case when eating out, as seen in the results presented for the 44 restaurants that serve an inexpensive menu. Portions in these restaurants are larger than necessary leading to calorie overeating and weight gain. Many of the ingredients in the dishes were over- and under-dosed, with over-dosing of starchy products, which provide macronutrients such as carbohydrates that are associated with weight gain (Rummo et al. 2023).(Rummo et al. 2023). This is because they are large glucose chains that are more easily assimilated by the body(Baig et al. 2023). In addition, due to their excitatory effects on the nervous system, they can reach an addictive level, which results in overeating without closing the cycle (Zainal Arifen et al. 2023).(Zainal Arifen et al. 2024).. On the other hand, a notable deficit was observed in the dosage of meat foods, both white and red meat, some offal, eggs and dairy products such as cheese. These foods mainly provide proteins of high biological value, which fulfil many functions in the cell, in the membrane and outside the cell(Chang et al. 2023)and are essential for life in the body(Guillamón Escudero et al. 2021).. Therefore, including protein in the diet is extremely important, even more so as the population currently consumes on average 0.5g/kg body weight (WHO n.d.).(WHO n.d.)which is a mega deficiency, as the minimum recommended intake is 0.8g/kg body weight (Chang et al. 2023).(Chang et al. 2023). Another group of foods that are deficient in the dosages of the most common preparations of the Puno diner are non-starchy vegetables, such as greens, including broccoli, cabbage, spinach, fruits such as tomatoes, and other vegetables such as onions and leeks. These foods, by providing mainly vitamins, minerals and dietary fibre, are involved in different processes that maintain a person's metabolic balance.(Chen et al. 2022). It is known that the daily requirement of vegetables should be 400g, which will provide the 30g of fibre required daily.(Meza-Ortiz, Martinez-Vazquez, and Yamamoto-Furusho 2022).and this must be provided by the foods ingested in the diet to maintain the colon in good condition, for digestive regulation and the healthy maintenance of the microbiota(Sánchez Almaraz et al. 2015)..

Thus, nowadays, budget menu restaurants are very popular due to the demand for fast food and the constant search for instant gratification. However, behind their affordable prices and the variety of dishes they offer, a worrying problem arises: excessive portions of rice, potato and noodles.(Miramontes-Escobar et al. 2020).. These colossal portions not only imply food waste, but can also negatively affect the health and well-being of diners. Excessive consumption of simple carbohydrates, present in large amounts of rice, potatoes and noodles, can raise blood sugar levels in the long term, which could lead to insulin resistance and type 2 diabetes(Ashcheulova et al. 2018).. In addition, these inordinate portions often lack adequate amounts of protein and vegetables, resulting in an unbalanced diet that is deficient in nutrients essential for optimal body function. Lack of protein affects muscle development(Chang et al. 2023)while a shortage of vegetables deprives the body of vitamins, minerals and fibre, crucial elements for digestive, immune and cardiovascular health (James and Wang 2019).(James and Wang 2019).

The consequences of this problem are not only limited to individual health, as excessive portions also generate food waste, which negatively impacts the environment(Carretero García 2018)..

CONCLUSION

The dosage of food in the breakfasts, lunches and dinners offered to consumers in the busiest areas of the city of Puno does not conform to the recommendations of the food dosage table proposed by the Peruvian National Centre for Food, Nutrition and Healthy Living. This is evidenced by the excesses in the group of cereals and tubers, deficiencies in the presence of vegetables, meat and fish, and the absence of fruits in the three meal times. Therefore, frequent attendance at these establishments that offer dishes high in calories, saturated fats, sodium and sugars, and low in fibre and essential nutrients at the different meal times would have an impact on the health of the population of Puno, increasing the risk of obesity, heart disease, type 2 diabetes and other chronic conditions.

REFERENCES

Ashcheulova, Tatiana, Ganna Demydenko, Tatiana Ambrosova, Kysylenko Kateryna, Nina Gerasimchuk, Oksana Kochubiei, Tatiana Ashcheulova, Ganna Demydenko, Tatiana Ambrosova, Kysylenko Kateryna, Nina Gerasimchuk, and Oksana Kochubiei. 2018. "Carbohydrate and Lipid Disorders and Adipokines Levels in Relation to Body Mass Index in Hypertensive Patients". *Revista Mexicana de Cardiología* 29(2):74-82.

Baig, J. A., I. G. Chandio, T. G. Kazi, H. I. Afridi, K. Akhtar, M. Junaid, S. Naher, S. A. Solangi, and N. A. Malghani. 2023. "Risk Assessment of Macronutrients and Minerals by Processed, Street, and Restaurant Traditional Pakistani Foods: A Case Study." *Biological Trace Element Research* 201(7):3553-66. doi: 10.1007/s12011-022-03429-7.

Carretero García, Ana. 2018. "Social, economic and environmental impacts arising from food loss and waste". *Przegląd Prawa Rolnego* (2(23)):127-39. doi: 10.14746/ppr.2018.23.2.9.

Chang, Liyang, Rongrong Tian, Zili Guo, Luchen He, Yanjuan Li, Yao Xu, and Hongmei Zhang. 2023. "Low-Protein Diet Supplemented with Inulin Lowers Protein-Bound Toxin Levels in Patients with Stage 3b-5 Chronic Kidney Disease: A Randomized Controlled Study." *Hospital Nutrition* 40(4):819-28. doi: 10.20960/nh.04643.

Chen, Man-Shuang, Don-Ying Wang, Hui-Yu Gong, Hui-Min Zhang, Jie Gao, and Song-Ping Luo. 2022. "The Association between Dietary Fiber and Infertility among US Women: The National Health and Nutrition Examination Survey, 2013-2018." *Hospital Nutrition* 39(6):1333-40. doi: 10.20960/nh.04056.

Guillamón Escudero, Carlos, José Miguel Soriano Del Castillo, Ángela Diago Galmés, José M. Tenías Burillo, and Julio Fernández Garrido. 2021. "Protein intake in community-dwelling postmenopausal women and its relationship with sarcopenia". *Nutricion Hospitalaria* 38(6):1209-16. doi: 10.20960/nh.03690.

Hernánzez Elizondo, Jessenia, Andrea Solera Herrera, Elizabeth Carpio Rivera, Alejandro Salicetti Fonseca, and David Hortigüela Alcalá. 2019. "Nutritional assessment and phytoestrogen exposure in a diet of students of the University of Costa Rica." *Nutricion Hospitalaria* 36(3):647-57. doi: 10.20960/nh.02109.

James, Armachius, and Yousheng Wang. 2019. "Characterization, health benefits and applications of fruits and vegetable probiotics". *CyTA - Journal of Food* 17(1):770-80. doi: 10.1080/19476337.2019.1652693.

Meza-Ortiz, Cinthya J., Sophia E. Martínez-Vázquez, and Jesús K. Yamamoto-Furusho. 2022. "Association of Dietary Fiber Consumption with Disease Activity in Ulcerative Colitis: An Exploratory Study in the Mexican Population." *Gaceta Medica De Mexico* 158(1):41-47. doi: 10.24875/GMM.M22000639.

Miramontes-Escobar, Herenia Adilene, Gladys América Prado-Guzmán, María de Jesús Toledo-Palomera, Jesús Enrique Báez-García, Sonia Guadalupe Sáyago-Ayerdi, Herenia Adilene Miramontes-Escobar, Gladys América Prado-Guzmán, María de Jesús Toledo-Palomera, Jesús Enrique Báez-García, and Sonia Guadalupe Sáyago-Ayerdi. 2020. "Nutritional profile according to socio-economic levels and menus provided in a soup kitchen in Mexico." *University and Health* 22(3):203-12. doi: 10.22267/rus.202203.192.

Omoniyi, Saheed Adewale, and Altine Cosmas. 2024. "Cooking practice and safety assessment of boiled cassava root sold in streets of Gashua, Yobe state Nigeria". *Food Chemistry Advances* 4:100562. doi: 10.1016/j.focha.2023.100562.

WHO. n. d. "Healthy eating". Retrieved 26 April 2024 (https://www.who.int/es/news-room/fact-sheets/detail/healthy-diet).

Redondo Del Río, María Paz, Beatriz De Mateo Silleras, Laura Carreño Enciso, José Manuel Marugán de Miguelsanz, Marina Fernández McPhee, and María Alicia Camina Martín. 2016. "Dietary intake and adherence to the Mediterranean diet in a group of university students as a function of sports practice." *Nutricion Hospitalaria* 33(5):583. doi: 10.20960/nh.583.

Rummo, P. E., T. Mijanovich, E. Wu, L. Heng, E. Hafeez, M. A. Bragg, S. A. Jones, B. C. Weitzman, and B. Elbel. 2023. "Menu Labeling and Calories Purchased in Restaurants in a US National Fast Food Chain." *JAMA Network Open* 6(12):E2346851. doi: 10.1001/jamanetworkopen.2023.46851.

Sánchez Almaraz, Rosalía, María Martín Fuentes, Samara Palma Milla, Bricia López Plaza, Laura M. Bermejo López, and Carmen Gómez Candela. 2015. "Indications of different types of fibre in different pathologies." *Nutrición Hospitalaria* 31(6):2372-83. doi: 10.3305/nh.2015.31.6.9023.

Singh, S., Soni, P. Lohani, A. Priya, A. Ranjan, and N. Nimavat. 2023. "Effect of Educational Intervention on Knowledge and Attitude about the Role of Vitamins, Minerals and Nutraceuticals in COVID-19 and Other Disorders among Medical and Nursing Undergraduates of a

Tertiary Care Teaching Hospital". *Clinical Nutrition ESPEN* 56:142-48. doi: 10.1016/j.clnesp.2023.05.004.

Villegas Villegas, Ricardo Yazmani. 2011. "Improvement of the manual dosing process in the production of balanced feed at the Balanfarina plant."

Zainal Arifen, Zainorain Natasha, Suzana Shahar, Kathy Trieu, Hazreen Abdul Majid, Mohd Fairulnizal Md Noh, and Hasnah Haron. 2024. "Individual and total sugar contents of street foods in Malaysia - Should we be concerned?" *Food Chemistry* 450:139288. doi: 10.1016/j.foodchem.2024.139288.

Zheng, Hong, Xinbin Chen, Xiaoling Bu, Xia Qiu, Demeng Zhang, Yitong Zhou, Junlong Lin, Jinghong Li, Wenjun Ma, and Ying Zheng. 2023. "Assessment of Dietary Nutrient Intake and Its Relationship to the Nutritional Status of Patients with Crohn's Disease in Guangdong Province of China." *Hospital Nutrition* 40(2):241-49. doi: 10.20960/nh.04395.

Buy your books fast and straightforward online - at one of world's fastest growing online book stores! Environmentally sound due to Print-on-Demand technologies.

Buy your books online at
www.morebooks.shop

Kaufen Sie Ihre Bücher schnell und unkompliziert online – auf einer der am schnellsten wachsenden Buchhandelsplattformen weltweit! Dank Print-On-Demand umwelt- und ressourcenschonend produziert.

Bücher schneller online kaufen
www.morebooks.shop

Printed by Books on Demand GmbH, Norderstedt / Germany